Energy Today and Tomorrow

Polly Goodman

New Hanover County Public Library
201 Chestnut Street
Wilmington, North Carolina 28401

Gareth Stevens Publishing

Please visit our website, www.garethstevens.com.
For a free color catalogue of all our high-quality books,
call toll free 1-800-542-2595 or fax 1-877-542-2596.

Library of Congress Cataloging-in-Publication Data

Goodman, Polly.
Energy today and tomorrow / Polly Goodman.
    p. cm. — (Earth alert!)
Includes index.
ISBN 978-1-4339-6002-4 (library binding)
1. Power resources—Juvenile literature. 2. Energy conservation—Juvenile
literature. I. Title.
TJ163.23.G66 2011
333.79—dc22
                              2010049258

This edition first published in 2012 by
Gareth Stevens Publishing
111 East 14th Street, Suite 349
New York, NY 10003

Copyright © 2012 Wayland/Gareth Stevens Publishing

Editorial Director: Kerri O'Donnell
Design Director: Haley Harasymiw

All rights reserved. No part of this book may be reproduced
in any form without permission from the publisher, except by a reviewer.

Printed in China

CPSIA compliance information. Batch WAS11GS. For further information contact Gareth Stevens, New York, New York at 1-800-542-2595

Picture acknowledgements
Shutterstock, cover picture; Axiom Photographic Agency (Jim Holmes) 20, (Jim Holmes) 28; James Davis Travel Photography 3; Ecoscene (Angela Hampton) 18, (Nick Hawkes) 25; Hodder Wayland Picture Library 6, 8 top, 12, 17 bottom, 27; Alexey Malashkevich/Shutterstock 1, (Tony Page) 5, (Jorn Stjerneklar) 8 bottom, (Homer Sykes) 9, (Yann Arthus Betrand) 10, (Peter Arkell) 11, (Charles Coates) 16, (Mark Henley) 19, (Javed A Jaferji) 22; W. Lord 26; Sheffield City Council 23 both; Splash Communications 21 both; (David Young Wolff) 4 top, (Mike Abrahams) 7, (Ken Graham) 13, (Tony Craddock) 17 top, (Dave Jacobs) 24; © Artiga Photo/Corbis 5; Istock 14.
Artwork by Peter Bull Art Studio.

Contents

What Is Energy?	4
Sources of Energy	6
Problems of Energy	12
How Much Energy	18
Using Less Energy	22
Renewable Energy	24
Glossary	30
Topic Web and Further Resources	31
Index	32

What Is Energy?

We need energy for everything we do. We use energy all the time and in many different forms.

When you were asleep last night, your body used energy to keep your heart pumping. When you got out of bed, the muscles in your legs used energy to let you walk. When you had breakfast, your toaster used energy to make bread into toast, and the refrigerator used energy to keep the milk cold. If you traveled to school by bus or car, your transportation used energy.

We use energy to make our bodies move and work. We also use energy for heating, lighting, cooking, transportation, and to run machines.

◑ Some things we use at home, such as irons, use energy from electricity.

Computers use energy from electricity. ➲

Damage from Energy

We use a huge amount of energy. The way we make and use some forms of energy can damage the environment. Many scientists believe we are in danger of running out of some energy sources, such as oil and gas.

We need to find ways of cutting down the amount of energy we use. We also need to develop forms of energy that do not harm the environment.

Smoke from an oil refinery pollutes the air with gases.

Activity

ENERGY DIARY

Keep an energy diary for a week. Each day, list everything you do and where the energy came from. Here are some examples.

TIME	ACTIVITY	TYPE OF ENERGY
07:30	Ate toast for breakfast	Electricity and my body
08:00	Bus to school	Diesel fuel

Sources of Energy?

RENEWABLE AND NONRENEWABLE

* The wind, the sun, and moving water are renewable sources of energy.
* Coal, oil, and gas are nonrenewable sources of energy.

Most of the earth's energy comes from the sun. Energy in the sun's rays helps plants to grow, and plants provide food for people and animals. Food is used to make our muscles work, which helps us survive.

The sun's energy is stored in wood, coal, oil, and gas. These are called fuels. We burn fuels to release their energy, which we use to drive machines and to make electricity.

Other sources of energy are the wind, the sun, and moving water.

Two farmers use energy from their bodies and from a machine to plant rice. ↻

Fuels

Wood is the oldest type of fuel. For centuries, it has been burned for cooking and heating. It is still used in poorer countries, where people cannot afford more expensive fuels.

Coal, oil, and gas are called fossil fuels. They were formed over millions of years from the fossilized remains of plants and animals.

Coal is made from the remains of plants, squashed together in layers. Oil and gas are formed from plants and animals that died in the sea. Fossil fuels have been used to power machines since the 1700s.

⌒ A coal miner deep underground.

Problems with Fossil Fuels

Fossil fuels are nonrenewable sources of energy because it would take millions of years to replace them. They also cause air pollution when they are burned.

Water

Water is used to make electricity, or hydropower. Dams built across rivers collect water in reservoirs. The water falls through turbines in the dam, which turn the energy into electricity.

⋂ An oil drill about to enter the ground.

Dams raise the water level in rivers, so the water falls farther and makes more energy. ⋃

TRUE STORY

OIL AND GAS UNDER THE CASPIAN SEA

Many fuel companies are working in the Caspian Sea because it holds vast amounts of oil and gas.

The Caspian Sea is a large sea in Central Asia. It is surrounded on all sides by Russia, Kazakhstan, Turkmenistan, Iran, and Azerbaijan.

It is hard and costly to reach the oil and gas. Companies need special equipment to drill, process, and transport the fuels. The cold weather adds to the difficulties. Yet the huge demand for oil and gas makes it worthwhile.

A group of international companies has set up the Kashagan project in the Kazakhstan part of the sea. Iran drilled its first well to explore for oil in 2010.

An oil rig at sea is often as big as a small town. ⮕

Nuclear Energy

Nuclear energy is made from uranium, which is a type of rock. The energy is made by splitting the nucleus of the uranium atom. Then power plants turn the energy into electricity.

Nuclear energy can be very dangerous if it is not carefully controlled. It gives off radioactivity, which damages all living things.

Inside a nuclear power plant. ⇒

ENERGY IN THE U.S.

In the United States, there are several main sources of energy: oil, gas, nuclear, and hydroelectric power. Wind and solar power are also used.

Electricity

Electricity is made from different energy sources, by machines called generators in power plants. Coal, gas, oil, and nuclear power plants use heat to make steam, which turns the generators. Water turns the generators in hydroelectric power plants.

Energy Around the World

Fossil fuels such as oil, gas, and coal are not found all around the world. Countries where they are found, such as the United States and the UK, have grown rich by using the fuels for transportation and industry. They have also made money by selling the fuels to other countries.

In some countries, many people cannot afford to buy fossil fuels. They still rely on animals to pull machines, and use wood for cooking and heating.

Some of these countries are finding other ways to make their own energy, such as hydropower. They are borrowing money to build hydropower plants.

Fossil fuels are used in industry, like this car factory.

Activity

ENERGY MAP

1. Trace a map of your state.
2. Draw symbols showing the location of power plants using information from your local library.
3. Show whether the power plants use coal, water, gas, nuclear energy, or wind to produce electricity.

Problems of Energy

Making and using energy can damage the environment in different ways. Some forms of energy are more harmful than others.

CARBON DIOXIDE

When fossil fuels are burned they release gases, especially carbon dioxide (CO_2). Here are the proportions of CO_2 emissions from different sources in the U.S.:

Power plants	34%
Transportation	33%
Homes	17%

Producing Electricity

When coal, oil, or gas power plants produce electricity, they release gases into the air, causing air pollution. Nuclear power plants also produce waste, which remains very dangerous for thousands of years. People are still looking for safe ways to get rid of nuclear waste.

An open coal mine in Germany.

Transporting Fuels

Fuels are carried huge distances, by trucks, tankers, and pipelines. These forms of transportation can damage the environment.

Engines in trucks and tankers burn even more fossil fuels and gases pollute the atmosphere. Oil leaks from tankers and pipelines kill wildlife and pollute the water. It can take years for a region to recover from an oil spill.

An oil pipeline crossing Alaska.

Running Out of Energy

Fossil fuels such as coal, oil, and gas are nonrenewable. Many people believe that if we continue using these fuels as much as we do today, they will start to run out in about 50 years' time. As they become more scarce, the prices will rise.

TRUE STORY

THREE GORGES DAM

The Three Gorges Dam, built on the Yangtze River in China, was completed in 2006. It is used to make hydropower. The dam's 26 turbines produce as much electricity as 15 coal-burning power plants. This prevents air pollution. It is intended to protect people from flooding as well.

Yet the dam was built at a huge cost to people and the environment. Around 1.2 million people living in nearly 500 cities, towns, and villages along the river lost their homes. Millions of plants and animals lost their habitat.

↻ The Three Gorges Dam, in China.

Some engineers did not think the dam should be built. They argued that a number of smaller and much cheaper dams on the Yangtze tributaries (small rivers flowing into the Yangtze) could have produced the same amount of power and prevented flooding, too.

Using Fossil Fuels

When engines in cars and other vehicles burn fossil fuels, they release gases into the atmosphere. These gases affect the world's climate. They also pollute the air and can form acid rain.

Global Warming

Most scientists believe that burning fossil fuels is making the world's climate warmer. The earth has a layer of gases around it. These are called greenhouse gases, which act like a blanket and keep the earth warm. Burning fossil fuels is increasing these gases in the atmosphere and making the world hotter.

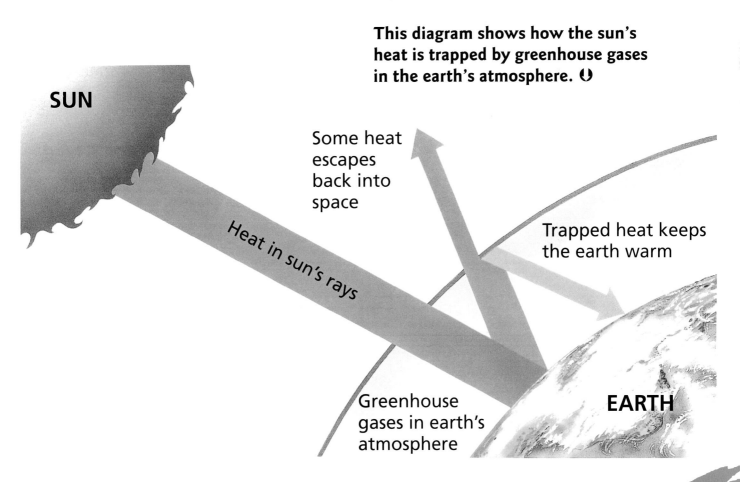

This diagram shows how the sun's heat is trapped by greenhouse gases in the earth's atmosphere.

The Effects of Global Warming

Most scientists agree that the earth could get warmer by another 3.2°F to 7.2°F (1.8°C to 4°C) by the end of this century. If the temperature rises by more than 3.6°F (2°C), huge areas of ice in the Arctic will melt and sea levels may rise by 7 to 23 inches (18 to 59 cm). Areas near coasts would be flooded and small islands would be completely covered in water.

A warmer climate would cause drought and water shortages. Crops could fail. This would cause famine and millions of people could be forced to move.

A warmer climate could mean rivers like this one would run dry. ↻

Mosquitoes that carry malaria are only found in warm climates. They could spread to new areas. Wildlife that could not adapt fast enough would become extinct.

Acid Rain

Acid rain is formed when gases from power plants and car engines are released into the atmosphere. The gases mix with water in clouds. When it rains, the rain is a weak acid mixture. It destroys trees and plants, and can kill wildlife in rivers.

This tree has been killed by acid rain.

Activity

Lichen can be green, gray, or brown.

AIR POLLUTION SURVEY

Lichen is a type of plant that grows on trees, rocks, and walls. It doesn't grow well in polluted air.

1. Look for lichen growing on a tree or a wall.
2. Find the place where it stops growing.
3. Look around to see what might be stopping it growing, for instance, pollution from traffic or factory chimneys.

How Much Energy

You can find out how much energy you use in your home or school by looking at a meter. Electricity, gas, and oil companies use meters to show how much we have used. Then they can calculate how much we have to pay. Energy is measured in joules.

Activity

SCHOOL ENERGY SURVEY

1. List everything in your school that uses energy and the type of energy it uses.
2. How does the energy come into the school buildings?
3. Where are the control switches?
4. Where are the meters and what do they show?

This woman uses energy in many different ways in her kitchen. How many machines can you see in the picture? ⇨

Lots of Energy

People in wealthy countries such as the United States and the UK use far more energy than people in poorer countries. People in rich countries have more cars, computers, and other machines. They have more money to buy the electricity and gas that run the machines. But some poorer countries, such as India and China, are catching up.

↷ A street in Calcutta, in India. More and more people are buying cars in developing countries like India.

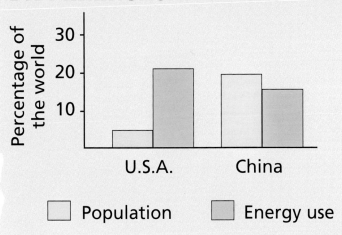

DIFFERENCES IN ENERGY USE

Not Enough Energy

In poorer, developing countries, many people still use wood for fuel. They cut down trees to use. As trees disappear, villagers have to go farther and farther to collect firewood.

The roots of trees help to keep the soil in place when it rains. In areas where most of the trees have been cut down, the soil is eroded, or washed away. This makes it harder to grow crops.

People are trying to find other sources of energy to use as fuel, such as animal dung.

◐ **A woman puts cow dung on sticks to use as fuel.**

TRUE STORY

CLOCKWORK RADIO

Most radios use energy from electricity or from batteries. Both of these energy sources can pollute the environment. In poor countries, some people cannot afford batteries.

Inventor Trevor Baylis wanted to make a radio that everyone could afford, including people in the remote countryside in Africa. He invented a clockwork radio.

When the radio is wound up, energy is stored in a metal spring. As the spring slowly unwinds, it releases its energy and powers a generator. The Freeplay clockwork radio takes about 20 seconds to wind up and lasts for up to an hour.

◯ The Freeplay clockwork radio.

◯ Clockwork radios can be used in places where there is no electricity.

Using Less Energy

If we want to make sure that fossil fuels do not run out, we have to cut down the amount of energy we use. There are many ways to do this.

These women only have wood as fuel for cooking. ➲

Wasting Less Energy

The simplest way of using less energy is not to waste any. Turn off lights when you leave a room. Switch off machines such as televisions and computers when they are not being used. Only use radiators in the rooms you are in. Install insulation in roofs and around windows—material that stops heat escaping.

Governments

Governments can encourage people to use less energy by making them pay taxes on the energy they use. They can also encourage people to use public transportation. This uses less energy per person than cars.

Governments can encourage energy companies to produce more energy from renewable sources, such as wind, water, and solar power.

SUSTAINABLE BLACON

The residents of the suburb of Blacon in Chester, UK, wanted to cut their fuel bills and help to improve the environment. In 2009, they set up Sustainable Blacon.

They have encouraged hundreds of households to insulate their homes and have created a green space for walking and cycling that provides a habitat for bats and bees. The group rents out bicycles to residents to promote cycling.

The group has made up two "Eco-houses" to show people how to save energy by making simple changes.

A volunteer works in the garden of the Eco-house.

A group of cyclists in Blacon prepare for a ride.

Renewable Energy

The world's population is growing. Developing countries are becoming industrialized. More and more energy is needed every day. Even if we find ways of cutting down the amount of energy we use at the moment, it is still important to develop renewable sources of energy to use instead of fossil fuels.

These solar dishes collect energy from the sun. ↻

◯ A tidal barrier and power plant across the estuary of the Rance River, in France.

Renewable Sources

Wind, wave, and tidal power can make electricity by using the flow of air, water, and tides to turn turbines.

Solar power traps the heat of the sun, and geothermal power uses the heat in rocks deep underground.

Other renewable energy sources include fuel from burning wood and methane from animal dung. Additional sources are crops such as corn, rapeseed, and sugar cane. These are called biofuels.

TRUE STORY

SOLAR-POWERED HOUSE

Bill and Debbi Lord live in Maine. Their house makes solar power.

The house has two types of solar panels. One type heats water. The water runs around the house in pipes, heating the radiators and providing hot water through the faucets. The other type makes electricity.

◯ The solar panels on this roof face the south, so they get sunshine all through the day.

A ventilation system lets out air but keeps in the heat. The roof and windows are insulated, so no heat is wasted.

At night, when there is no sunshine, Bill and Debbie use electricity from the electric company. During the day, the house provides more electricity than they need, so the electric company buys it from them.

The Problems of Renewable Energy

Renewable energy sources do not release harmful gases into the atmosphere. But they can harm habitats when they are built and can change the landscape.

When the barrage of a tidal power plant is built across a river, it disturbs the river's wildlife. Wind turbines often cover a large area of land and some people think they spoil landscapes. Growing crops for biofuels reduces the amount of land for growing food. Renewable energy can also be less reliable than energy from fossil fuels. For instance, wind is always needed to drive wind turbines.

Wind power is one of the cleanest forms of renewable energy. ↻

RENEWABLE ENERGY USE

In 2009, 10.6 percent of electricity in the U.S. came from renewable energy sources. Some states aim to double this amount by 2020. In 2010, Germany produced 16 percent of electricity from renewables. It aims for a complete switch to renewables by 2050.

The Future

We have found out that renewable sources of energy are less harmful to the environment than fossil fuels and that they will not run out. But it may take some time for renewable energy sources to completely replace fossil fuels.

Many developing countries are already building power plants that use renewable energy sources, such as hydropower. Wealthy countries, such as the United States and the UK, are also choosing to use more renewable energy sources. This will protect the environment for people in the future.

Solar panels outside a hospital in Tanzania.

Activity

MAKE A SOLAR PANEL

1. Line the inside of a cardboard box with tinfoil.
2. Cut a piece of PVC tube, three times the length of the box.
3. Set the tube into the box in a curvy shape, using wire to attach each bend.
4. Put a lump of modeling clay in one end of the tube and pour water in the other end.
5. Fill the tube with water. Check the water temperature. Put the solar panel in a sunny place and leave it for a few hours.
6. Test the temperature of the water to see if it has become warmer.

How do you think the tinfoil helped heat the tube?

In which countries would solar power work best?

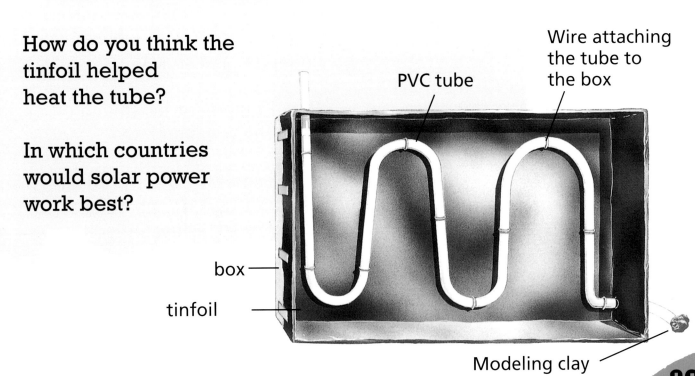

29

Glossary

Acid rain Rain that contains pollution from factories and traffic.

Atoms The tiny particles that make up every chemical.

Barrage An artificial barrier in a river.

Carbon dioxide A gas breathed out by people and animals or produced by burning carbon.

Developing countries Poorer countries where most people work on farms and industries are small.

Emissions Gases that are sent out into the air.

Environment Everything in our surroundings, such as the earth, air, and water.

Fossil fuels Coal, oil, and gas, formed over time from the remains of plants and animals.

Hydropower Energy made from moving water. It is also called hydroelectric power.

Industrialized Countries with large industries using machinery.

Malaria A serious disease caused by the bite of some types of mosquito.

Nonrenewable Sources of energy that cannot be replaced as quickly as they are used up. Coal, oil, and gas are nonrenewable sources of energy.

Pollution The process of making air, water, or soil dirty.

Renewable Able to be replaced. The wind, the sun, and moving water are renewable sources of energy because they will not run out.

Reservoirs Places where water is collected and stored.

Rig Equipment for taking oil or gas from the ground or sea.

Solar power Power produced from the sun's rays. It can be used to heat water or to produce electricity.

Turbines Engines or motors that are made to work by the power of water, steam, or air.

Further Information

Topic Web

MUSIC
- Compose a song or rap to give a message about how people use energy

GEOGRAPHY
- Human use of energy around the world
- Extraction of resources
- Mapwork
- Environmental issues: e.g. global climate and sea level change, land use, land/water/air pollution

HISTORY
- Water power and mills
- Invention of steam power
- Energy/pollution

ARTS & CRAFTS
- Using energy as a stimulus for painting/modeling
- Design a poster on reducing energy use

DESIGN AND TECHNOLOGY
- Simple machines and mechanisms to demonstrate energy transfer

MATH
- Measuring of units of energy
- Simple statistics
- Graphs about energy use

SCIENCE
- Formation of fossil fuels
- Nuclear reactions and radioactivity
- Electricity
- Environmental issues: e.g. types of renewable energy

ENGLISH
- Using energy as a stimulus for creative writing
- Reports/letters
- Library skills

Books

Energy Revolution: Building a Green Community by Ellen Rodger (Crabtree, 2008)

Biofuels: Sustainable Energy in the 21st Century by Paula Johanson (Rosen Classroom, 2010)

Acid Rain (Our Planet in Peril) by Louise Petheram (Capstone Press, 2000)

Doable Renewables: 16 Alternative Energy Projects for Young Scientists by Mike Rigsby (Chicago Review Press, 2010)

Websites

Energy Kids
http://www.eia.doe.gov/kids/
Find out all about energy: its history and sources, the conservation of energy, and games.

Energy Quest
http://energyquest.ca.gov/index.html
Discover exciting facts about all forms of energy and energy conservation! Includes movies and pictures.

Environmental Kids Club
http://www.epa.gov/kids/
Learn all about the environment, plants, and animals.

Index

acid rain 15, 17
Arctic 16

batteries 21
biofuels 25, 27

Caspian Sea 9
China 14, 19
climate change 15, 16
clockwork radios 21
coal 6, 7, 10, 11, 12, 13
computers 4, 19, 22
cooking 4, 7, 11, 22

dams 8, 14

electricity 4, 5, 8, 10, 11, 14, 18, 19, 21, 25, 26
engines 15
environment 5, 12, 13, 28

food 6, 16, 27
fossil fuels 7, 8, 11, 12, 13, 15, 22, 24, 27, 28
France 25

gas 5, 6, 7, 9, 10, 11, 12, 13, 18, 19

generators 10, 21
geothermal power 25
Germany 12, 28
global warming 15, 16
greenhouse gases 15

heating 4, 7, 11, 26
hydroelectric power (hydropower) 8, 10, 11, 14, 23, 25, 28

India 19
insulation 22, 23, 26
Iran 9

machines 4, 6, 7, 10, 11, 18, 19, 22

nuclear power 10, 11, 12

oil 5, 6, 7, 8, 9, 10, 11, 12, 13, 18

pipelines 13
pollution 5, 8, 12, 13, 14, 15, 17, 21
power plants 10, 11, 12, 13, 17, 25, 27

radioactivity 10
renewable energy 6, 23, 24, 25, 26, 27, 28

solar power 10, 23, 24, 25, 26, 28, 29
sun 6, 15, 24, 25
Sustainable Blacon 23

Three Gorges Dam 14
tidal power 25, 27
transportation 4, 9, 11, 12, 13, 23
turbines 8, 14, 25, 27

UK 10, 11, 19, 23, 28
United States 11, 19, 26, 28
uranium 10

water power (see hydropower)
wind power 6, 10, 11, 23, 25, 27
wood 6, 7, 11, 20, 22, 25

32